AF881058

SOTHIS PRESS

Sothis Press aims to completely reimagine the public domain book with an emphasis on universal access to captivatingly designed print editions. We believe that technology will enable every single book ever published to be perpetually accessible in beautiful print form. Sothis is beauty in ubiquity.

www.sothis.press
1 Waterhouse Square, London, ECIN 2ST

Handbook to the Matlick Tellurian

Guide to Mathematical and
Astronomical Geography

Isaac Newton Matlick

SOTHIS PRESS

ISBN 978-1-0368-0813-6

Handbook to the Matlick Tellurian
is in the public domain.

This edition is published by
Sothis Press © 2023. The company
is registered in England and Wales.
The Company Number is 14900389.

ISAAC NEWTON MATLICK, A. M.

ANSI

HAND BOOK

TO THE

Matlick Tellurian

GUIDE TO MATHEMATICAL AND ASTRONOMICAL GEOGRAPHY

BY

ISAAC NEWTON MATLICK, A. M.

PUBLISHED BY

AMERICAN TELLURIAN MFG. CO.

SEATTLE CABLEGRAM "TELLURIAN" WASH.

AGENCIES IN PRINCIPAL CITIES IN U. S.
AND IN FOREIGN COUNTRIES

Press of
MECHANICS PUBLISHING COMPANY
Seattle, Wash., U. S. A.

Isaac Newton Matlick, A. M.

Isaac Newton Matlick, inventor of the tellurian which bears his name, was born in Pleasant Township, West Virginia, in the year 1846; and died in San Francisco, California, February the third, 1913: His remains being interred in Lone Fir Cemetery, Portland, Oregon. It was in this city where Prof. Matlick served many years as a principal of schools that he brought to completion the invention of his tellurian. To this great task he gave thirty-one years of painstaking effort. The achievement should give him rank among the great mathematicians and mechanics of the world.

Those who knew him testify to his great strength of character, combined with kindliness of heart and fine humility. His invention, wonderful in its capacity to make plain the great laws of the world and the solar system, should make his name familiar to the youth of many generations.

PSALM XIX.

The Heavens declare the glory of God
And the firmament showeth His handiwork.
Day unto day uttereth speech
And night unto night showeth knowledge.
There is no speech nor language;
Their voice is not heard;
Their line is gone out through all the Earth
And their words to the end of the World.
In them hath He set a tabernacle for the Sun
Which is as a bridegroom coming out of his
chamber
And rejoiceth as a strong man to run His
course.
His going forth is from the end of the
Heavens,
And His circuit unto the ends of it,
And there is nothing hid from the heat
thereof.

Psalms XIX, vs. 1-6.

Table of Contents

List of Illustrations

Diagrams

Introduction

This Hand-book is designed as a guide to the Teacher or Student in using Matlick's New Tellurian. In the preparation of this work we have endeavored to present the principal points to be shown by the Tellurian, in a plain, simple manner, so that the inexperienced Teacher, as well as the Pupil, may readily comprehend them. There are many other points that the thoughtful Teacher will find valuable to present which can be accurately explained with the Tellurian.

The Tellurian represents accurately, in the most simple manner, the intricate and complex movements of the Earth and Moon around the Sun, showing their relative positions to each other any day of the year, so that the following points may be readily comprehended:

1. The causes of a change of season.
2. The causes of day and night, and of the variations in length of day and night in different latitudes.
3. The portion of the Earth illuminated each day in the year.

4. The point on the Earth where the direct rays of the Sun will shine each day of the year.
5. The cause of the difference in length of days or nights.
6. The inclination of the Earth's equator and axis to the plane of the ecliptic.
7. The direction of the Sun's rays to the surface of the Earth at different seasons of the year, and at any hour in the day.
8. Causes of the present shape of the Earth.
9. The nature and causes of trade winds.
10. The Sun's declinations.
11. Causes and phenomena of the Equinoxes.
12. Nutations of the Earth's polar axis.
13. The difference between solar and siderial time.
14. The difference between siderial and clock time.
15. Causes of the Sun being slow or fast.
16. How mean time is computed.
17. The extent and duration of twilight on any part of the Earth.
18. The different phases of the Moon.
19. Causes of annular, partial and total eclipses of the Sun.
20. Eclipses of the Moon.

21. Umbra and penumbra.
22. The Moon's nodes and their revolution.
23. Revolution of the apsides.
24. Causes and phenomena of the tides-- spring and neap tides.
25. Effects of perigee and apogee upon eclipses and tides.
26. The siderial and lunar month.
27. Comparative size of the Earth and Moon.
28. Center of gravity of the Earth and Moon.
29. Causes of the harvest Moon.

The following geographical and astronomical terms can also be illustrated, viz.: Latitude, longitude, meridians, palallels, axis, poles. tropics, polar circles, zenith and nadir, vertical circles, celestial meridian, prime vertical, azimuth, etc.

In this Hand-book the phenomena of the tides are explained on a different hypothesis from that generally given in text-books; a comprehension of the lunar motion of the Earth, which is represented by the Tellurian, being necessary to explain the subject.

Thus it will be readily seen that it is not only a necessity, but an indispensable apparatus for the Public School, and valuable for High Schools and Colleges, where the more intricate problems are to be studied, as it is impossible for any Teacher, however apt in illustration, or concise in language, to present clearly to the mind of the pupil the above points without the aid of proper apparatus. As such apparatus, it is conceded by the leading educators that the Matlick Tellurian has no equal.

Although heretofore, for the want of proper apparatus, these subjects have been imperfectly understood, it is now certain that, in the immediate future, they will be properly explained in every school-room through the use of this marvelously complete and accurate instrument.

Description of the Tellurian

The top of the stand (Plate 1) is elliptical to represent the ellipticity of the Earth's orbit. The periphery is made to represent the zodiacal belt—a belt 16° in width with the ecliptic as its center.

Upon the center of the stand is a revolving hub having a socket (No. 7) into which is placed the Earth-arm, (No. 3) which is held in place by a set-screw (No. 8). At the outer end of the Earth-arm, driven by a bevel gear, is a vertical shaft which by a simple mechanism carries the globe made to represent the Earth, and also the Moon-arm-cam (No. 10). Upon the Moon-arm (No. 4) is placed the Moon which is constructed upon the same scale as the Earth. The Earth is supplied with a time band (No. 13), a day-band (No. 16), and an adjustable meridian (No. 17).

As the vertical shaft is rotated the Earth and Moon revolve about their common center of gravity, the Earth maintaining its proper inclination to the plane of the ecliptic, the North Pole being always directed to the same objective point, thus showing very clearly the third or lunar motion of the Earth. The Moon rises and falls in its orbit and passes through all its phases, showing the inclination of its orbit to the ecliptic and the gyration of its nodes, which are so arranged that not only the cause but the date of recurring eclipses may be shown by the Tellurian.

The day-band shows the exact portion of the Earth illuminated by the Sun each day in the year. With the day-band, time-band and adjustable meridian a child of ten can tell the time of sunrise and of sunset for any place and date. The degree of latitude on the Earth immediately under the noon point of the time-band will indicate the distance of the vertical rays of the Sun north or south of the Equator for each day in the year.

The mechanism may be thrown out of gear by releasing the clutch (No. 12) by means of the thumb-screw. When released the Earth may be revolved on its orbit without rotating on its axis; or the Earth and Moon may be revolved about their common center of gravity without moving the Earth along its orbit. The axis rod (No. 14) is rotated by simple friction so that the Earth may be rotated on its axis without moving any other parts.

The top and periphery of the stand is covered with a colored chart. On the central disk is the general plan of the Solar System, including all the Planets, their satellites, and a comet, with their corresponding sizes and distances, as near as practicable. The perihelion point of each Planet is marked in its orbit with the letter P. The Planets all move in the same general direction around the Sun as that of the Earth. On the main portion of the stand, and extending to the edge, the chart shows the months of the year, and their divisions into days; the different signs of the zodiac through which the Sun will pass any month or day; the right ascension of the Sun in hours and degrees for any day; the number of minutes the Sun is slow or fast for any day in the year; the autumnal and vernal equinoxes; the summer and winter solstices; the aphelion and perihelion; when the seasons begin, etc. The chart on the rim shows the ecliptic or orbit of the Earth through the great zodiacal belt, showing the signs in which the Earth will appear, looking from the Sun, and its right ascension in hours and degrees for any day in the year; the astronomical and almanac signs representing each particular sign of the zodiac. At the perihelion point the Earth is shown to be several inches nearer the Sun than at the aphelion.

The instrument is so perfect and durable in its construction that nothing less than actual violence can break of put it out of order, and, with the accompanying instructions, any one can operate it, and understand the subjects designed to be illustrated.

The object of this highly instructive apparatus is to illustrate and simplify, to the eye of the learner, the whole theory of Celestial Mechanics, including the rudiments of Astronomy and Mathematical and Astronomical Geography.

No other instrument can *claim* but a small part of what this instrument *actually does*, as the Matlick New Tellurian is the only instrument that correctly represents the true motions of the Earth and Moon, showing their exact relation to each other and to the Sun for each day in the year. There are no fewer than one hundred theorems and problems included within the above subjects that this instrument is capable of illustrating.

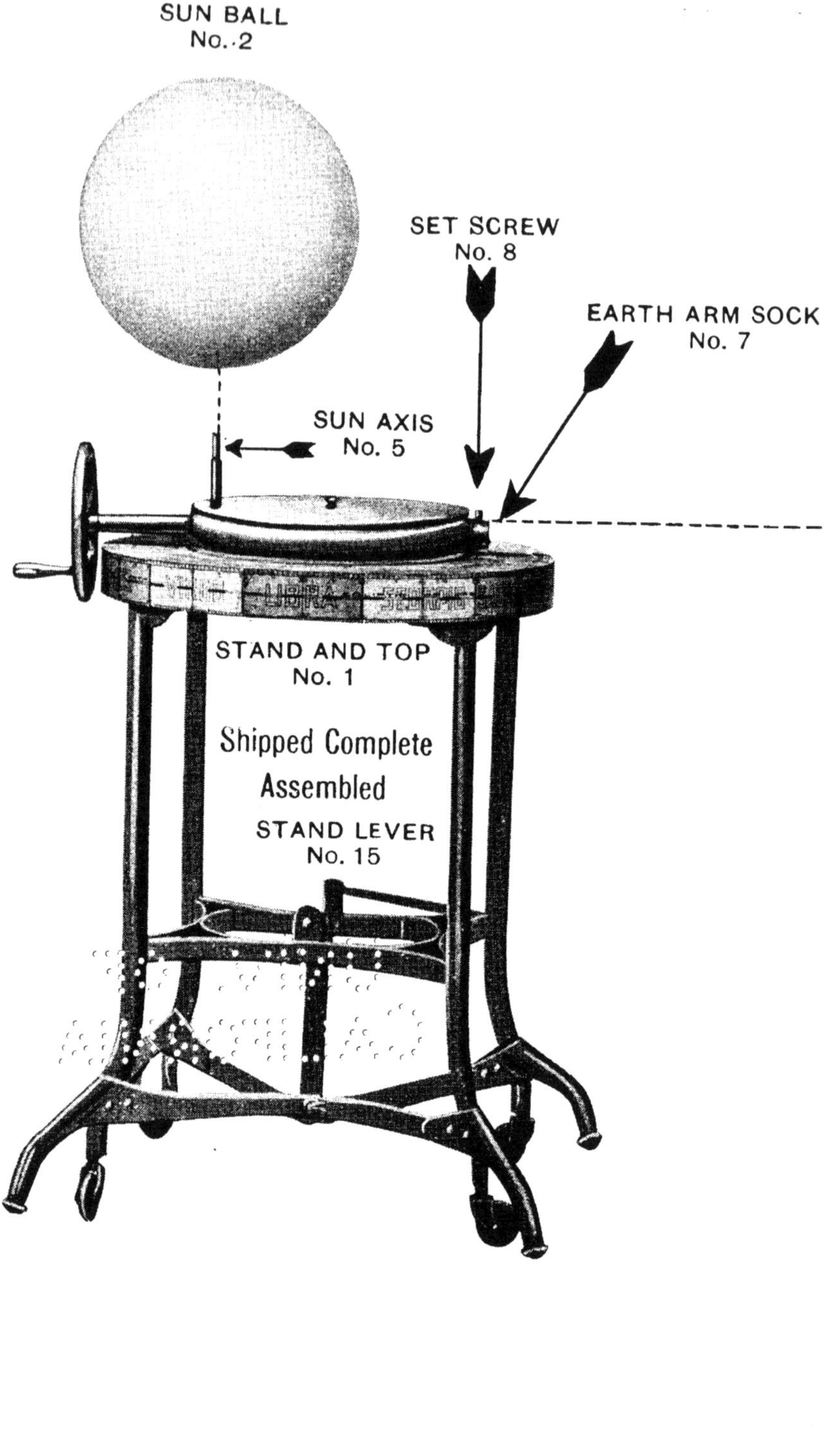
SUN BALL
No. 2
SET SCREW
No. 8
EARTH ARM SOCK
No. 7
SUN AXIS
No. 5
STAND AND TOP
No. 1
Shipped Complete
Assembled
STAND LEVER
No. 15

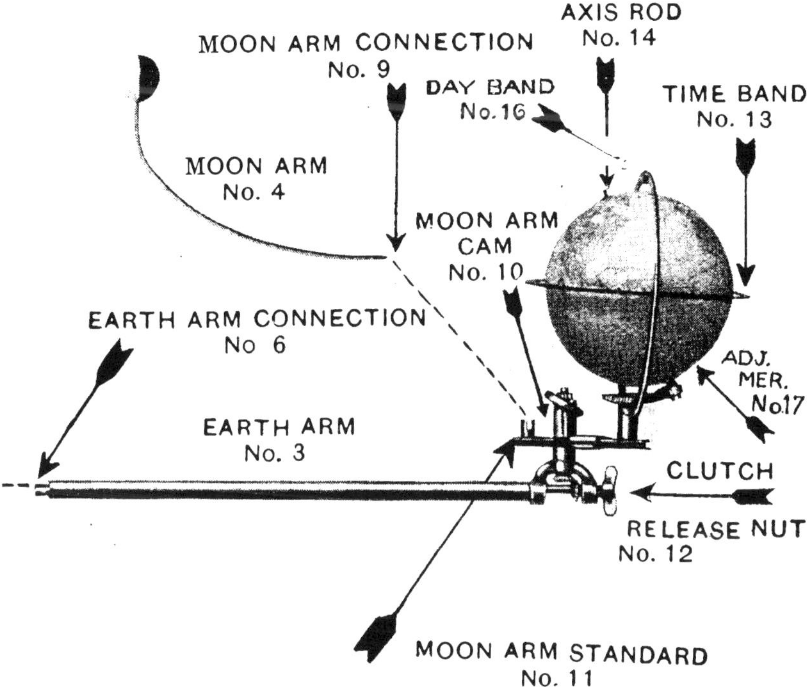
AXIS ROD
No. 14
MOON ARM CONNECTION
No. 9
DAY BAND
No. 16
TIME BAND
No. 13
MOON ARM
No. 4
MOON ARM
CAM
No. 10
EARTH ARM CONNECTION
No 6
ADJ.
MER.
No.17
EARTH ARM
No. 3
CLUTCH
RELEASE NUT
No. 12
MOON ARM STANDARD
No. 11

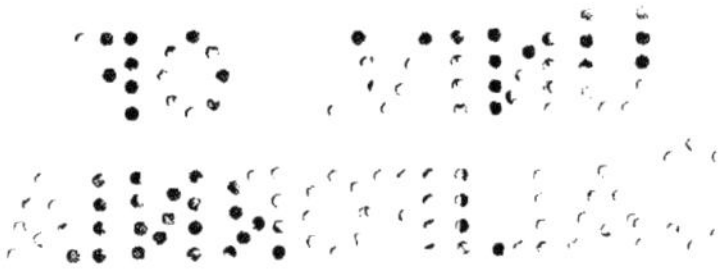

The Earth

The form of the Earth is very nearly that of a globe, slightly flattened at the poles. This figure is known to mathematicians as oblate spheroid. This flattening at the poles is very slight, being about 1-300th part of the diameter of the Earth. The axial diameter is about twenty-six miles less than the equatorial. The circumference of a great circle of the earth is about 25,000 miles, and the diameter of that circle is about 8,000 miles. The surface of the Earth embraces an area of 200,000,000 square miles, of which 50,000,000 is land and the remainder water. For the convenience of determining positions on the Earth, certain imaginary lines, or circles, are supposed to be drawn upon it.

The Axis of the Earth is the diameter about which it revolves, and the extremities of the axis are donominated the poles; the one above the equator the North Pole, and the one below, the South Pole.

The Equator is a great circle, at equal distances from the poles, and dividing the Earth into two parts, called the Northern and Southern Hemispheres.

The Parallels of Latitude are small circles parallel to the Equator, and are drawn for every ten degrees, and are numbered, from the Equator to the poles, 90°., North of the Equator is called North Latitude, and south of the Equator is called South Latitude. The width of a degree of latitude is 69 1-6 miles.

Meridians are great circles passing from the North Pole to the South Pole, and crossing the Equator at right angles. Upon the globe used on the Tellurian to represent the Earth, the meridians are marked every 15°, which corresponds to one hour of time. That is, it will require one hour for the vertical rays of the Sun to pass over 15° of longitude. These circles are numbered to the east 180° and to the west 180°, from the meridian of Greenwich, which passes near London. East of this established meridian is called East Longitude, and west, West Longitude. The width of a degree of longitude on the Equator is 69 1-6 miles, but terminates at the poles, where the width is 0.

The Tropics. It will be observed, in the motion of the Tellurian, that the vertical rays of the Sun fall upon the Earth 23½° north and 23½° south of the Equator, when the Earth is in different points of its orbit, marking, respectively, the Tropic of Cancer and the Tropic of Capricorn. The day circle of the Earth, at the point where the vertical rays are farthest north, will show that the rays of the Sun shine 23½° beyond the North Pole, and fall 23½° short of reaching the South Pole, or when the vertical rays are farthest south the reverse of this order will be the result. These positions of the Sun's rays to the Earth mark the Arctic Circle, 23½° from the North Pole, and the Antarctic Circle, 23½° from the South Pole.

Zones. We thus see why the Earth is naturally divided into five zones. The region north and south of the Equator, and between the Tropics of Cancer and Capricorn, over which pass the vertical rays of the Sun, is called the Torrid Zone, and is 47° in width. The region between the Tropic of Cancer and the Arctic Circle is called the North Temperate Zone, 43° in width. The region between the Tropic of Capricorn and the Antarctic Circle, the South Temperate Zone, 43° in width. The regions beyond the polar circles are called the North and South Frigid Zones.

Daily Motion: If we observe the positions of the Sun, Moon and Stars for a few successive nights, we shall see their relative positions gradually change. In our observations for any day or night, all the heavenly bodies appear to move to the west. The motions are, of course, only apparent, as there is absolute proof of the motions of the Earth in an opposite direction to the apparent motion of the heavenly bodies. All the apparent motion of the heavenly bodies that seemingly pass to the westward, is the result of the daily revolution of the Earth on its axis to the east. As a result of the diurnal motion of the Earth, we see the Sun rise in the east and set in the west each day, producing day and night. When the Sun passes below the horizon its light is no longer visible, and one-half the Earth is wrapped in darkness. As the pin in the center of the solar semi-circle band indicates noon on any meridian brought to it, so the meridian on the opposite side of the Earth must indicate mid-night; consequently to an observer on the midnight meridian, the Earth is directly between him and the Sun. Strictly speaking, the Earth revolves on its axis in about 23 hours and 56 minutes, but, as the Earth is continually changing its longitude, the day is 24 hours long. It will, therefore, make 366 revolutions in a year of 365 days.

Difference in the Length of Day and Night.

Since the Sun shines on but one-half of the Earth at a time, it is evident that at the Equator the days and nights must always be of equal length. If the axis of the Earth were perpendicular to the plane of the ecliptic, so that the Sun always shone from pole to pole, the days and nights would forever be of the same length on all parts of the Earth. But to an observer, on any parallel north or south of the Equator, the length of the days and nights will vary in different parts of the year, and the nearer he approaches the poles the greater will be the difference, until he reaches the poles, where the days and nights are alternately six months in length. Sometimes the days are longest, and again the nights. These differences are regular and uniform, and their recurrence is the same for each year.

Annual Motion. The various changes in the length of day and night must be looked for from another cause than the rotation of the Earth on its axis. In observing the fixed Stars, it will be noticed that they do not appear on the same meridian at the same time each night, but about four minutes earlier, so that they appear to have a general motion to the west, in addition to their apparent diurnal motion. Any particular Star seen to rise in the east at a certain hour in the night will be seen to gradually course its way across the heavens so that in six months from that night, at a corresponding hour, it will fall below the western horizon. It will, of course, rise and set each day or night, but will come to the meridian earlier each night. From this and other causes it has been fully demonstrated that the Earth has an annual course around the Sun from the west to east. By this motion the Sun appears to be carried around the ecliptic through the fixed Stars. Now, as the Sun appears high in the heavens in one part of the ecliptic, and again falls toward the horizon, it is evident that the Earth's axis is not perpendicular to the plane of the ecliptic, but inclines from this perpendicular. It is found that the Sun's highest and lowest meridian altitude differ 47°. Now, one-half of this, or 23½°, must represent the inclination of the Earth's axis, and the Earth's Equator must necessarily make an angle with the plane of the ecliptic of 23½°.

The Poles. The North Pole of the Earth is denominated the elevated pole, because it is always about 66½° above the plane of the ecliptic, or about 23½° from a perpendicular to the plane of the ecliptic; and the South Pole is denominated the depressed pole, because it is about 66½° below the plane of the ecliptic, and 23½° from a perpendicular to the plane. These motions and the direction of the Earth's axis are so perfectly represented in the Tellurian that the student cannot fail to comprehend the entire subject.

The Day Circle on the globe, as used on the Tellurian, when properly adjusted, shows exactly the portion of the Earth illuminated each day, and the variation in the lengths of day. So the learner's attention is directed wholly to it at the present. At the Vernal Equinox the Sun's rays shine from pole to pole, but, as the Earth moves along in its orbit, the North Pole is gradually turned into the sunlight, and the South Pole into the darkness, until the Earth arrives at the summer solstice —June 21—when the North Frigid Zone is wholly in the sunlight, and the South in darkness. It will be seen that for six months the days in the Northern Hemisphere have gradually increased in length, and in the Southern Hemisphere have diminished. They are now longest north of the Equator and shortest south. Now, as the Earth is moved around to the autumnal equinox, the days naturally diminish in length in the Northern Hemisphere, and increase in the Southern Hemisphere, until they are again equal. If the motion is continued to the winter solstice—December 21—the days in the Northern Hemisphere will be shortest, and in the Southern Hemisphere longest. The North Frigid Zone is wholly in darkness, and the South Frigid Zone in the sunlight. For any day between these dates the day circle will show the corresponding lengths of day and night on any portion of the Earth. The length of a parallel on the Earth shown by the day circle to be illuminated, as compared by that portion on the opposite side, will indicate their relative lengths. The meridians marked on the Earth, followed to the horizontal band with the 24 hours of day and night on the dark and illuminated sides, will serve as a means of measurement.

When the Sun Rises, and When It is Midnight at the Poles. When the Sun passes the vernal equinox, it rises to the Arctic, or elevated, Pole, and sets to the Antarctic Pole. When the Sun arrives at the summer solstice it is noon at the North Pole, and midnight at the South Pole. When the Sun passes the autumnal equinox, it sets to the North Pole, and rises to the South Pole. When the Sun arrives at the Winter solstice, it is midnight at the North Pole, and noon at the South Pole; and when the Sun comes again to the vernal equinox, it closes the day at the South Pole, and lights up the morning at the North Pole.

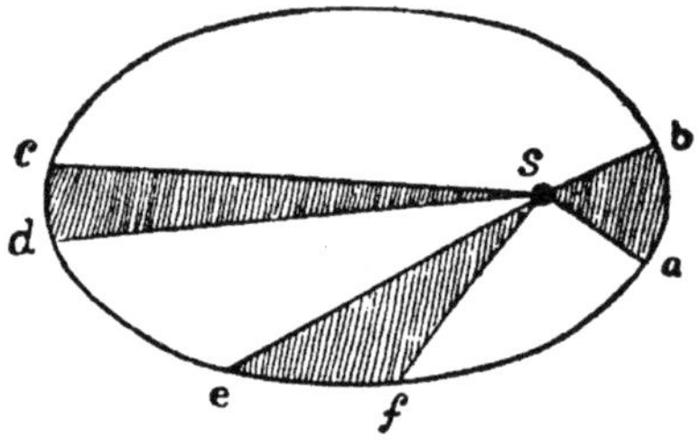

—*Illustrating Kepler's Second Law.*
Fig. 1.

The illuminating band, or day circle, on the Earth, will show how the sunlight approaches or recedes from the pole.

The Nights at the Poles Not Equal. By consulting an almanac for any year one discovers that the time required for the Sun to make its apparent journey from the autumnal equinox to the vernal equinox is 178½ days, while from the vernal equinox to the autumnal requires 186½ days. This results from what is known as Kepler's second law of planetary motion. (Fig. 1). The radius vector describes equal areas in equal times. That is, in our winter, the Earth being in perihelion, moves more rapidly in its orbit than when in aphelion. This is clearly exhibited by the Telurian.

Night Within the Polar Circle. At the Arctic Circle, 23° 27½′ from the Pole, the longest day is 24 hours, and goes on increasing as you approach the Pole. In latitude 67° 18′ it is 30 days; in latitude 69° 30′ it is 60 days, etc. The same takes place between the Antarctic Circle and the South Pole, with the exception that the day in the same latitude south is a little shorter, since the Sun is not so long south of the Equator as north of it. At Spitzbergen the day gradually increases in length from the first glimpse of the Sun on February 21, to 12 hours on March 21; then to 24 hours on April 21, when the Sun remains continuous above the horizon until August 21. The day then alternates with night, decreasing from 24 hours to a parting glimpse on October 21, when night continnes for four months. In the southern part of Nova Zembla we find continuous day and night of about six weeks each, and then day and night alternate for twenty weeks each.

Further north, the periods of alternate day and night are shorter, decreasing to the Pole. At Wanderbus, in Norway, the day lasts from the 21st of May to the 22d of July, without interruption.

Day in the Temperate and Torrid Zones.

The greatest length of day in the Torrid Zone, which must be on the tropics, is 13½ hours. The greatest length of day in the Temperate Zone, which must be on the polar circle, is 24 hours. At Portland, Oregon, the longest day has 15½ hours; at Boston, 15¼; at Berlin and London, 16½; at Stockholm and Upsaal, 16½; at Hamburg, Dantzic and Stettin, 17, and the shortest, 7. At St. Petersburg and Tobolsk the longest day has 19, and the shortest 5 hours. At Bornea, in Finland, the longest day has 21½, and the shortest 2½ hours. The night must, in all cases, be 24 hours, minus the length of the day, and vice versa. As the change of the Sun's declination is less at the solstices than at the equinoxes, so the change in the relative length of day and night must also be variable in the same places. To illustrate this point more fully, the learner's attention is called to the position of the Earth to its orbit. It will be noticed that at the equinoxes the Earth moves in a direction very nearly indicated by its axis; so that from March 21 to April 21 the Sun moves northward about 10°, and from April 21 to May 21 about 9°, and from May 21 to June 21 about 4°; so that the Sun changes its declination very slowly at the solstices, as the Earth then is moving very nearly in a direction indicated by a parallel of the Earth. The same cause somewhat affects the apparent diurnal motion of the Sun to and from the noon point.

It follows that during the year every portion of the Earth must have an equal amount of day and night; that is, there must be in all parts six months day and six months night. In these estimates no account is taken of the refraction of the atmosphere, which increases the length of the day, by making the Sun appear more elevated above the horizon than it really is.

To Determine When the Sun Rises and Sets.

By observing the circle, the time the Sun will rise or set can be determined, as well as the length of day or night. To determine when the Sun will rise or set at any particular point, mark where the parallel of the given point intersects the day band, and trace the meridian directly to where it intersects the horizontal band marked with the hour of the day or night. This point of intersection will show the exact time of the rising and setting of the Sun.

From this the length of day and night can be determined easily for any day in the year.

To find when the Sun rises and sets in any region where the day or night is more than 24 hours long, move the arm around and notice the date that that portion of the Earth appears on the illuminated side of the day circle, which will give the date of the Sun's appearance, and the portion that passes on the dark side, for the disappearance of the Sun.

The Earth's Orbit and the Zodiacal Belt.

The Earth's orbit around the Sun is elliptical, as shown by the top of the stand of the Tellurian. This is exaggerated, or is more elongated than if drawn to a true scale, but it serves to bring more forcibly to the eye the ellipticity of the Earth's orbit. The Sun occupies one of the foci of the ellipse, so that the Earth, in its course around the Sun, is brought very much nearer to the Sun in one point of its orbit than in another. The Earth is in perihelion January 1, when it is 91,000,000 miles from the Sun, and in aphelion July 1, when it is about 94,000,000 miles distant, being about 3,000,000 miles nearer the Sun in one part of its orbit than another. These points are plainly shown by the Tellurian.

The great belt of the heavens through which the Earth and Sun appear to travel is called the zodiac. This belt extends about 8° on each side of the ecliptic, and is shown by the chart on the rim of the stand. The orbits of all the major Planets in the solar system lie within this belt. The Sun appears to travel through this belt as the Earth makes its annual journey around the ecliptic. Commencing with the position of the Sun at the vernal equinox, the early astronomer divided the ecliptic into twelve signs, of 30° each, and afterwards gave the following names to them: Aries, Taurus, Gemini, Cancer, Leo, Virgo, Libra, Scorpio, Sagittarius, Capricornus, Aquarius and Pisces. At first, or about 2,200 years ago, the constellations of Stars corresponded with the sign in name, but, owing to the precession of the equinoxes, which will be explained hereafter, the constellations and signs have entirely changed, so that the constellation Pisces occupies the sign Aries, and the constellation Aries the sign Taurus.

When the Sun appears to us to be in one part of the ecliptic, the Earth, as seen from the Sun, appears in the point diametrically opposite. Thus, when the Sun appears in the vernal equinox at the first point of Aries, the Earth is actually in the opposite equinox at Libra. This may be observed by looking on the top plate on the top of the stand for the sign in which the Sun appears, and on the rim for the sign in which the Earth will be at a given time.

We have no historic record of this division of the zodiac into signs, and the ideas of the authors can only be inferred from collateral circumstances. It has been fancied that the names were suggested by the seasons, the agricultural operations, and so on. Thus, the spring signs—Aries the Ram, Taurus the Bull, and Gemini the Twins—are supposed to mark the bringing forth of young by the flocks and herds. Cancer, the Crab, marks the time when the Sun, having attained its greatest declination begins to go back toward the Equator; and the Crab having been supposed to move backwards, his name was given to this sign. Leo, the Lion, symbolizes the fierce heat of summer. Virgo, the Virgin, gleaning corn, symbolizes the harvest. In Libra, the Balances, the day and night balance each other, being of equal length. Scorpio, the Scorpion, is supposed to have marked the presence of venomous reptiles in October, while Sagittarius, the Archer, symbolizes the season of hunting. The explanation of Capricornus, the Goat, is more fanciful, if possible, than that of Cancer. It was supposed that the animal, ascending the hill as he feeds, in order to reach the grass more easily, on reaching the top, turns back again, so that his name was used to mark the sign in which the Sun, from going south, begins to return to the north. Aquarius, the Water Bearer, symbolizes the winter rains, and Pisces, the Fishes, the season of fishes.

Change of Seasons

Effects of Heat and Cold—Their Causes.

All the perceptible changes of the seasons are brought about by the difference in the temperature of the atmosphere. The green grass of spring, the golden grain of summer, the seared leaves of autumn, and the drifting snows of winter, are the results of heat and cold. The next question which arises in the mind of the learner is, what causes these various temperatures at the different times of the year, and on different portions of the globe? As the internal fires of the Earth have little or no effect at its surface, it is evident that our only source of heat is the Sun. The intensity of heat depends mainly upon the direction with which the Sun's rays fall upon the Earth. It has been demonstrated that as much heat is produced upon the same area from a vertical Sun in eight hours as would be produced at an agle of 60° in sixteen hours. Whether the rays fall on a particular point of the Earth vertically or obliquely depends upon the relative position of the Earth to the Sun. An additional cause of the different temperature is found in the fact that when the Sun is above the horizon at any place, at that place the Earth is receiving heat; and when the Sun is below the horizon, it is parting with its heat, by a process which is called radiation. The quantities of heat thus received and imparted in the course of the year must balance each other at every place, or the equilibrium of temperature would not be supported. Whenever the Sun remains more than twelve hours above the horizon of any place, the general temperature of that place will be above the mean state; when the reverse takes place, the temperature, for the same reason, will be below the mean state. The continuance of the Sun above the horizon of any place depends entirely upon his declination or altitude at noon, and this also determines the angle at which the rays fall upon the Earth at that point. Thus, the two causes of heat necessarily act at the same time, for the rays of the Sun are always most vertical when it shines longest above the horizon. The axis of the Earth is inclined to the plane of its orbit, and thus it is that in the course of its annual revolution around the Sun the rays fall at different angles on every portion of the globe.

The manner in which the Earth changes its relative position to the Sun during its yearly revolution is perfectly represented by the Tellurian.

Winter

Place the Earth and Moon in position, as previously explained and shown in Fig. 2. At this point, December 21, the Northern Hemisphere will have its greatest inclination from the Sun, and the rays of the Sun will fall more obliquely upon this portion of the Earth, and will be less effective in producing heat, and in the Southern Hemisphere the rays will fall more nearly vertical, consequently more heat will be the result. Thus, while the Northern Hemisphere has winter, the Southern Hemisphere will have summer. The Earth is now at the winter solstice.

The Arctic Circles and Tropic of Capricorn.

At this point the rays of the Sun will not reach the North Pole by 23½°, and will shine 23½° beyond or under the South Pole marking the Arctic and the Antarctic Circles. The vertical rays of the Sun will fall 23½° south of the Equator, and mark the Tropic of Capricorn; this marks the beginning of winter in the Northern Hemisphere, and the beginning of summer in the Southern Hemisphere. The Earth now enters the sign of Cancer, and the Sun the sign of Capricorn. It will now be con-

Figs. 2 and 6.

Fig. 3.

tinual night at the North Pole and continual day at the South Pole. At this time the Sun does not shine half way around the Earth in the Northern Hemisphere, and on more than half the Southern Hemisphere; that is, the Sun will shine on less than 180° of a given parallel in the Northern Hemisphere, and on more than 180° of a given parallel in the Southern Hemisphere. A few days later, or the 1st of January, the Earth is in perihelion, or nearest to the sun.

Spring

Pass the Earth around so that the long rod is directly over March 20, the vernal equinox. At this point the Earth neither inclines to nor from the Sun, and the days and nights are equal in all parts of the Earth, as the Sun shines from Pole to Pole, and the vertical rays fall directly upon the Equator. This marks the beginning of spring in the Northern Hemisphere, and the beginning of autumn in the Southern Hemisphere. The vertical rays are traveling northward, bringing warmth and heat into the Northern Hemisphere, and gradually receding from and leaving it cold and bleak in the Southern Hemisphere. At this point the Earth enters the sign of Libra, and the Sun the sign of Aries.

Summer

Continue the movement of the Earth around to June 21, the summer solstice, at which point the Northern Hemisphere of the Earth will lean toward the Sun, and the rays will strike more directly upon the portion of the Earth north of the Equator, and obliquely upon the Southern Hemisphere. This marks the beginning of summer in the Northern Hemisphere, and the beginning of winter in the Southern Hemisphere. The rays of the Sun now fall vertically upon the Earth $23\frac{1}{2}°$ north of the Equator, and mark the Tropic of Cancer, and also shine $23\frac{1}{2}°$ beyond the North Pole. (Fig. 3.)

The Tropic of Cancer and the Arctic Circle. At the Tropic of Cancer the vertical rays have reached their farthest point north of the Equator, and from there start south again. At this point continual day exists in the North Frigid Zone, and continual night in the South Frigid Zone. The Earth at this time enters the sign of Capricorn and the Sun the sign of Cancer. A few days later, or July 1, the Earth is in aphelion, or farthest from the Sun.

Autumn

Move the Earth around to September 22, the autumnal equinox, which marks the beginning of autumn in the Northern Hemisphere, and spring in the Southern Hemisphere, at which point the Earth again neither inclines to nor from the Sun, and the days and nights are equal throughout the Earth. The Earth now enters the sign of Aries, and the Sun Libra. By continuing the movement of the Earth until it again arrives at December 21, it will have completed its yearly revolution, showing the proper positions of the Earth and Moon in relation to the Sun every day during the year; remembering that the Earth should revolve on its axis every day, and that the North Pole of the Earth should at all time point in the same direction.

The True Cause of Temperature. Now, as the Sun's rays fall most obliquely when the days are shortest, and most directly when the days are longest, these two causes, namely, the duration and intensity of the solor heat, together, produce the temperature of the different seasons. The reason why we have not the hottest temperature when the days are longest, and the coldest temperature when the days are shortest, but in each case about a month afterwards, appears to be that a body once heated does not grow cold instantaneously, but gradually; and so of the contrary. Hence, as long as more heat comes from the Sun by day than is lost by night, the heat will increase, and vice versa.

Our Winter When Nearest the Sun. It may seem strange to the learner that we have our winter when nearest the Sun, and our Summer when most distant; but it must be remembered that the temperature of any particular part of the Earth is not so much affected by the distance of the Sun as by the directness or obliqity of its rays. Hence, though we are farther from the Sun on the 21st of June than on the 21st of December, yet as the North Pole of the Earth is turned more directly into the light at that time, so that the Sun's rays strike its surface less obliquely than in December, we have a higher temperature at that period, though at a greater distance from the Sun.

Twilight

When the Sun is below the horizon, the rays passing through the upper portion of the atmosphere are reflected and refracted by the molecules of the air, so that we are enabled to see objects by this reflected light, called twilight, at the dawn and close of day, some time before the appearance, and after the disappearance, of the direct rays of the Sun. The length or duration of twilight depends on many conditions. The atmosphere extends upward about 80 miles, and in cold regions the reflection and refraction is much greater than in warmer regions. Generally twilight extends from 15° to 18° beyond the portion illuminated by the direct rays. The duration at the Equator is about one hour, or a few minutes more at certain times, as when the Sun is at the solstice. In higher latitudes, or further north or south, the duration is much greater. Stockholm has a period of twilight lasting from sunset to sunrise for a period of four months. On the 85th parallel twilight lasts for twenty-eight days. These periods are, of course, when the Sun has its greatest declination north. The same would be noticed in corresponding latitudes south when the Sun has its declinations south. The day circle on the Earth, as used in the Tellurian, will fully illustrate these subjects. As twilight extends about 15° to 18° beyond this band, its duration can be fully determined by observing the time required for any particular point 15° to 18° beyond this circle to be brought into the direct rays in the diurnal and annular motions of the Earth.

The Moon

With the exception of the Sun, our interest in the Moon is greater than in that of any other heavenly body, as it is by far the nearest to us. Its mean distance from the Earth is about 240,000 miles, but, owing to the ellipticity of its orbit and the attractive force of the Sun, it varies from 10,000 to 20,000 miles upon each side of this mean during each monthly revolution, with an average oscillation on each side of 13,000 miles. The greatest possible distance is 259,600 miles, and the least distance is 221,000 miles from the Earth; but it rarely approaches these limits. The conditions required to produce this great oscillation are the Earth to be in a perihelion and the Moon in apogee, and in conjunction, for the greater distance; and the Moon in perigree and in op position, for the least distance. The Moon's diameter is 2,160 miles, or a little less than two-seventh that of the Earth; its volume is, therefore, about one-fiftieth that of the Earth, but owing to the greater density of the Earth, its mass is only about one-eightieth.

The Axial Revolution of the Moon. One of the most remarkable features of the Moon is,

Plate III.
The Moon.

that it revolves once on its axis while making one complete lunation, consequently presents the same face to the Earth continually, so that no human eye has ever seen but one side of the Moon. The lunar day is 29½ times as long as the terrestial day. The Sun will, therefore, shine on the Moon's Equator for nearly fifteen of our days, and will not be seen again for the same period. As a result the changes of temperature from the lunar day to the lunar night are very great. The heat of the day would perhaps be far greater than that known anywhere on our globe, while the excessive cold of our arctic winter would hardly equal that of the lunar night.

On the Moon. To an observer on one side of the Moon the Earth would appear like an immense Moon passing through all of the phases, but would never set; while if he were on the other side of the Moon this terrestrial sphere would never be visible. The light and dark observed upon the Moon is owing to the unequal reflection of light caused by the great diversity of her surface. By viewing the Moon through a telescope, immense mountains and deep crevices and craters may be seen, but no water ar atmosphere is at all discernable; and as we know life, it could not exist on that dead body. By the most careful determination yet made, it is found that the Sun gives 619,000 times as much light as a full Moon; but, while this is the comparison in light it has been carefully computed that the Sun only gives 82,600 times as much heat. This discrepancy is doubtless owing to the fact that the Moon has been raised to such a high temperature by the Sun's rays falling upon her surface continually for so long a time, that the Moon would thus reflect much of her heat.

The Effects of the Attractive Force of the Moon. The greatest effect of the Moon's attractive force, so far as yet understood, has been explained in the phenomena of the tides. There is also a tide in the atmosphere, produced from the same influence. There is no evidence that the Moon affects the Earth, or its inhabitants, in any other way than by its attraction. It is not improbable, however, that, owing to the peculiar relation of the Sun and Moon at times, and their declination north and south, producing wonderful changes in atmospherical tides, that storms and various changes in the weather would be the result. A thorough investigation of these causes and effects would doubtless do much to establish the true law of storms.

Phases of the Moon. To explain with the Tellurian, look in an almanac and determine when new Moon will occur the nearest to January 1 and briug the long arm over that date.

Fig. 4.

Fig. 5.

Throw the mechanism "out of gear" and bring the Moon around so that it is over the long arm between the Earth and Sun. This will show new Moon. Throw in gear. In 1909 new Moon occurred on January 21 and if the Tellurian is adjusted for that date and the long arm moved around it will show all the phases of the Moon for that year, and for 1910 it will show new Moon on January 11, thus illustrating accurately the phases of the Moon for any year or number of years and the dates of their occurrence. Keep the light side of the Moon toward the Sun in explaining the phases of the Moon. The Moon, being an opaque body, shines by the reflected light of the Sun. At this point the light of the Moon cannot be seen from the Earth, and is, therefore, said to be new Moon, or in conjunction. (Fig. 4.) Move the rod forward a short distance in the proper direction, and a small crescent will be observed. Continue the movement one-fourth of the distance in its orbit, and one-half the light side will be seen, which is the first quarter of the Moon or quadrature. (Fig. 5.) Again, move the bodies forward until the Moon is directly opposite the Earth from the Sun. The whole face of the Moon will now be seen, and it is full Moon or opposition. (Fig. 6.) Continue the movement one-fourth farther, and the Moon will arrive at third quarter, and the opposite half of the light side of the Moon will be seen from that which was shown at first quarter. It is again in quadrature. (Fig. 7.) The bodies may now be moved forward until the Moon is again in conjunction, remembering that the observer all the time is looking from the Earth beneath the Moon.

The Movement of the Moon. The time required for the Moon to move from new Moon to new Moon is 29½ days, while it will be observed that it will move completely around the earth in 27 1-3 days. The difference in these periods is occasioned by the onward movement of the Earth in its orbit. The movement of the Moon eastward will also explain the cause of the Moon rising on an average fifty minutes later each day.

Apogee and Perigee. In the movement of the Moon in its orbit, it will be observed that the Earth and Moon approach and recede from each other. The nearest approach of these bodies is called perigee, and the farthest point apogee. These two points in the Moon's orbit are also called the apsis, and the line connecting them the line of the apsides.

Declination of the Moon. The declination of the Moon north or south, or high Moon or low Moon, depends on two things: The in-

Fig. 7.

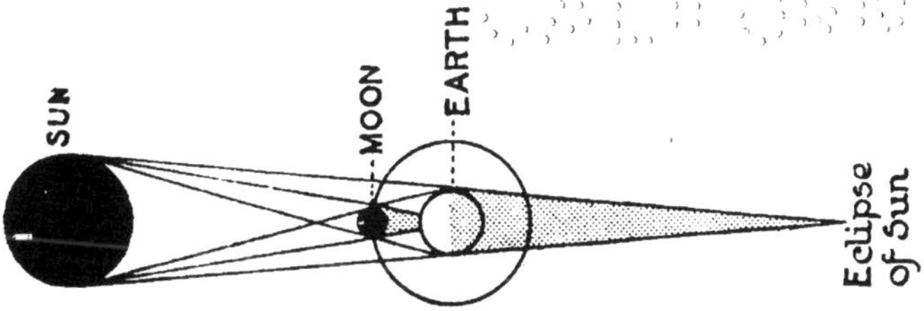

Fig. 8.

clination of the Moon's orbit to the plane of the ecliptic, which is about 5°, and the inclination of the Earth to the plane of the ecliptic, but mostly the latter, which will be observed by the Tellurian. The dim face of the whole disk of the Moon, when only a part of the illuminated side is seen, is due to the reflection of the light from the Earth on the Moon.

Eclipses of the Sun and Moon

An eclipes of the Sun and Moon has always been observed with great interest, and their recurrence terrified the ancients, and even the superstitutious of modern times. A close observation of the prenomena soon revealed the fact that an eclipse of the Sun always took place at new Moon, and of the Moon at full Moon, so that the motions of these bodies could not be watched long without the cause being seen. It was evident that if the Moon should pass between the Earth and Sun, the illuminating rays would be obscured, and thus we find the early Astronomers were well acquainted with the eclipses and their cause.

General. Cause of an Eclipse. If the Moon moved on the same plane with the Earth, it is evident that an eclipse would occur at every new and full Moon, but as the orbit of the Moon inclines to the plane of the ecliptic, about 5°, the Moon will generally pass above or below the Sun, and consequently there will be no eclipse. (Fig. 5.) The points where the Moon crosses the ecliptic are called the Moon's nodes. If the Moon is in the neighborhood of one of its nodes at new or full Moon, there will be an eclipse. (Fig. 2.)

Solar Eclipses—Their Nature and Causes. The number and nature of eclipses that occur each year depend upon the relation of the nodes to the Sun at the time of new and full Moon. To illustrate: On August 24, 1877, the Sun passed the Moon's descending node, but the times of new Moons nearest to that date were August 8 and September 6. At the first date the Moon passed above or north of the ecliptic, so that only a partial eclipse was visible in the northern part of Siberia, while at the latter date the Moon had passed so far south that only a small eclipse was visible at Cape Horn. Thus, there were two solar eclipses while the Sun was passing the one node, but very small. One year later, 1878, the Sun passed the node about August 4, and the new Moon occurred on July 29, the Moon being so close to the node that a total eclipse was visible. Every time the Sun passes a node there must be an eclipse, and as it passes both nodes each year, it is evident that there must be two solar eclipses each year, and it is possible for four to occur to some parts of the earth's surface. (Fig. 7)

Lunar Eclipses. It was noted by the earliest Astronomers that the Earth cast a shadow, and that as the Moon passed into this shadow it become eclipsed. That lunar eclipses only occur occasionally depends upon the same general principles of solar eclipses. The Earth's shadow, like the Sun, is in the plane of the ecliptic. Owing to the great magnitude of the Sun, the Earth's shadow is much smaller at the point where the Moon passes it than the Earth. (Fig. 8.) The Moon must therefore be very near one of its nodes at full Moon, or it will fail to strike the shadow, and will either pass above or below it. It is, therefore, evi-

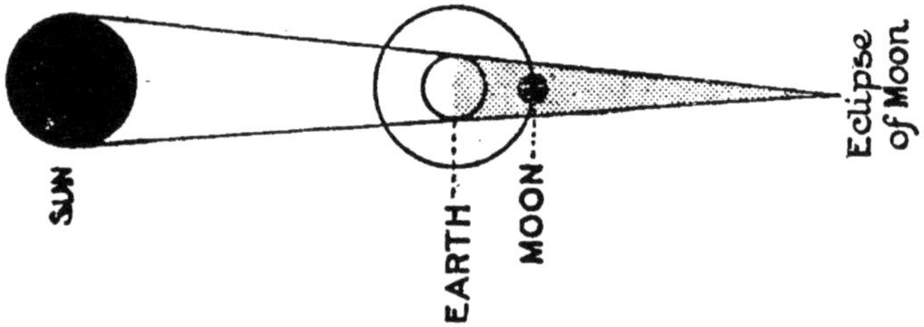

Fig. 9

dent that lunar eclipses are of less frequent occurrence than solar eclipses. A whole year may pass without there being a lunar eclipse, and there never can be more than two.

Total, Annular and Partial. The nature of an eclipse varies with the relative position of the Earth, Sun and Moon. If new Moon occurs when the Sun is at or very near the node, and the Earth is in the region of aphelion, and the Moon is in the region of perigee, the angular diameter of the Moon will exceed that of the Sun, and the shadow of the Moon will then fall upon the Earth, and the whole disk of the Sun will be hid; this is called a *total eclipse.* If a similar eclipse should take place when the Moon is in the region of apogee, and the Earth in the neighborhood of perihelion, the angular diameter of the Sun will exceed that of the Moon, and the Moon's shadow will not reach the Earth, then only a circular portion of the Sun will be hid and a ring of light around the edge of the Sun will be visible; this is called an *annular eclipse.* If the Moon does not pass centrally over the Sun, but a little above or below the ecliptic, so that only a part of the Sun is hid, it is said to be a *partial eclipse.* So with lunar eclipses; if the shadow only strikes a portion of the Moon, it is a partial eclipse, but if the whole disk of the Moon is immersed in the shadow, it is a total eclipse. As the shadow of the Earth where the Moon crosses it is larger than the Moon, a lunar eclipse cannot be annular.

The Sun, and the Moon's Nodes. There are two periods in each year when the Sun passes the Moon's nodes, and, therefore, eclipses will occur during those seasons. A solar eclipse may occur eighteen days before and after the Sun's passage through the node, while that for lunar eclipses extends 11½ days each side of the node, making a total season for solar eclipses of 3 days, and 23 days for lunar eclipses, each time the Sun passes a node.

The Changes of the Moon's Nodes. The Moon's nodes are continually changing or falling back to meet the Moon, so that eclipses will occur on an average of about twenty days earlier each year. To find the middle of an eclipse season, or the time the Sun passes a node for twenty-five or thirty years, subtract 19 2-3 days from any of the dates given herein for the middle of the eclipse period for each subsequent year. We find for 1881 that the Sun was in the node about June 6, and, as full Moon occurred June 11, there was a total eclipse of the Moon on that date. The Sun passed the other node about November 25, 1881, and the new Moon occurred on November 21; there was an annular eclipse of the Sun at that date. On May 17, 1882, the Sun was in the node again, and as the Moon was new on that day, a total eclipse occurred; but, as new Moon occurred at night, or when America was on the opposite side of the Earth from the Sun, it was invisible in this country. The falling back of the Moon nodes may be clearly illustrated with the Tellurian by placing the nodal points marked on the inclined disk in line with the Earth and Sun for any date, and then, with all the parts operating, move the long arm nearly around the Sun or within 20 days of the starting point, and the nodal points on the inclined disk will be found in the same line or relative position toward the Sun as at the starting point.

The Saros. The Moon will return to the same node in 27.2 days and is called the "nodal revolution of the Moon," or Draconic month. The other period of 29.5 days is called the "synodical revolution" of the Moon. Now, the curious consequence of these figures shows that there are 242 Draconic months, 233 Lunations (New Moons) and 19 returns of the Sun to one and the same node of the Moon at nearly the same time, and all are accomplished in 18 years, 10 days and a few hours. That is, if the Sun and Moon should start together from the same node, they would be found together very near the same node in the period above mentioned, and eclipses would occur almost, though not quite, in the same regular order in about 18 years, 10 days and 7 hours. This is the celebrated Chaldean "Saros," and was used by the ancients for the prediction of the eclipses alike of the Sun and Moon, and may be used by the modern as an interesting and instructive pastime.

The mechanism of the Tellurian is so timed that it produces these results.

The effects of the operation of the "Saros" may be noticed in the following two noteworthy

"Saros" groups of solar eclipses during the second half of the nineteenth century:

1842	...........July 8	1850	.....August 7
1860	..........July 18	1868	.....August 17
1878	...........July 29	1886	.....August 29
1896	.......August 9	1904	..September 9

For a more complete elucidation of the above facts, we give in tabular form all the eclipses of a succession of half of a "Saros," or nine years, and thus show more clearly the principles we have endeavored to explain.

	Eclipses	Approximate Mid-interval.
1894—	March 21, eclipse, Moon.... April 6, eclipse, Sun......	.....March 29*
	September 15, eclipse, Moon. September 29, eclipse, Sun..	September 22**
1895—	March 11, eclipse, Moon.... March 26, eclipse, Sun......	.....March 18*
	August 20, eclipse, Sun..... September 4, eclipse, Moon.. September 18, eclipse, Sun..	..September 4**
1896—	February 13, eclipse, Sun.. February 28, eclipse, Moon.	..February 20*
	August 9, eclipse, Sun..... August 23, eclipse, Moon....	...August 16**
1897—	February 1, eclipse, Sun...	....February 1*
	July 29, eclipse, Sun......	......July 29**

1898—January 7, eclipse, Moon... January 22, eclipse, Sun....	....January 14*
July 3, eclipse, Moon...... July 18, eclipse, Sun.......	.....July 10**
1898—December 13, eclipse, Sun.. December 27, eclipse, Moon. 1899—January 11, eclipse, Sun....	..December 27*
June 8, eclipse, Sun....... June 23, eclipse, Moon.....	.....June 15**
December 2, eclipse, Sun... December 16, eclipse, Moon.	...December 9*
1900—May 28, eclipse, Sun...... June 13, eclipse, Moon.....	..... June 5**
November 22, eclipse, Sun..	.November 22*
1901—May 3, eclipse, Moon...... May 18, eclipse, Sun......	......May 10**
October 27, eclipse, Moon.. November 11, eclipse, Sun..	...November 3*
1902—April 8, eclipse, Sun....... April 22, eclipse, Moon..... May 7, eclipse, Sun........	.....April 22**
October 17, eclipse, Moon... October 31, eclipse, Sun....	....October 24*

One * denotes the ascending node, and two ** denotes the descending node. We give here some recent eclipses:

1904—March 17, annual eclipse of the Sun**

September 9, total eclipse of the Sun.

1906—February 9, total eclipse of the Moon.
February 23, partial eclipse of the Sun.
July 21, partial eclipse of the Sun.*
August 4, total eclipse of the Moon.**
August 19-20, partial eclipse of the Sun.*
1909—June 3, total eclipse of the Moon.**
June 17, partial eclipse of the Moon.
November 27, total eclipse of the Moon.
December 12, partial eclipse of Sun.**
1910—May 9, total eclipse of the Sun.*
May 23-24, total eclipse of the Moon.**
November 2, partial eclipse of the Sun.**
November 16, total eclipse of the Moon.

One * denotes the ascending node, and two ** denote the descending node.

*To Adjust the Moon's Nodes to Show the Eclipses With the Tellurian...*Throw "out of gear." The inclined disk on which the Moon's support travels causing the Moon to rise and fall in its orbit, has two points marked upon it half way between the highest and lowest points. These points are intended to indicate the Moon's node; that is, when the small traveler that supports the Moon is at these points, the Moon is in its node. The one where the Moon is rising is called the ascending node, and the other, where the Moon is passing downward, is called the descending node. Determine from an almanac, astronomy or appended table of the eclipses the time of an

Plato IV.
Eclipse of the Sun.

eclipse of the Sun or Moon, and whether at the ascending or descending node, and bring the node in a line with the long arm, the arm being over the date indicated for the eclipse of the Sun. The Moon should be in a line between the Earth and Sun, or new Moon. If a total eclipse, the Moon will be in a line with the node, and between the center of the Earth and the center of the Sun, or in the plane of the ecliptic. If a partial eclipse, and visible in the Northern Hemisphere, the nodal point should be moved back a little so the Moon will be partially above the plane of the ecliptic. If the eclipse is visible in the Southern Hemisphere, the nodal points should be moved forward a little so the Moon would be partially below the plane of the ecliptic. These same conditions apply to the eclipses of the Moon, only the Moon should be in opposition, or full Moon, where the shadow of the Earth would fall upon it. Having thus adjusted the mechanism and "thrown in gear" the operation of the Tellurian will readily show the *cause* and *dates* of the eclipse of the Sun and Moon for a number of years.

Third Motion of the Earth and Lunar Orbit. It is in consequence of the *mutual* gravitation of all the several parts of matter, which the Newtonian law supposes, and the *third law, that action is equal to reaction, and in oppo-*

site direction, that the Earth and the Moon revolve about their common center of gravity in their monthly orbit, and continue to circulate, without parting company, in a greater annual orbit around the sun. We may conceive this motion by connecting two unequal balls by a short stick, which *to their common center of gravity* is suspended by a long string and made to gyrate conically round a point vertically below that of suspension. Their joint system will circulate as one pendulous mass about this imaginary center, while yet they may go on circulating round each other in subordinate gyrations, as if the stick were quite free from any such tie, and merely hurled through the air.

If the Earth alone, and not the Moon, gravitated to the Sun, it would be dragged away and leave the Moon behind, and vice versa; but, acting on both, they continue together under its attraction, just as the loose parts of the Earth's surface continue to rest upon it. It is, then, in strictness not the Earth or the Moon which describes an ellipse around the Sun, but their common center of gravity. The effect is to produce a small, but very perceptible monthly equation in the Sun's apparent motion as seen from the Earth, which is always taken into account in calculating the Sun's place. The Moon's actual path in its

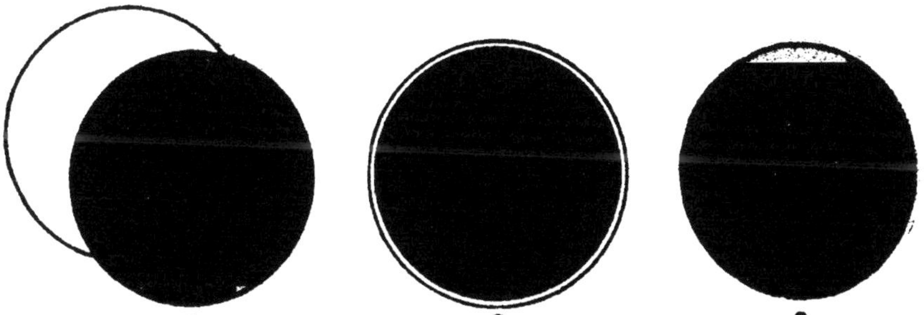

1 Partial. 2 Annular. 3 Tota l.
Fig. 10.

compound orbit around the Earth and Sun is an epicycloidal curve, intersecting the orbit of the Earth twice every lunar month, and alternately within and without it. But as there are not more than twelve such months in the year, and as the total departure of the Moon from it either way does not exceed 1-400th part of the radius, this amounts only to a slight undulation upon the Earth's ellipse, so slight, indeed, that if drawn in true proportion, on a large sheet of paper, no eye, unaided by measurement with compasses, would detect it. The real orbit of the Moon is everywhere concave towards the Sun.

The motions of the Earth and Moon about their common center of gravity are so perfectly represented in the Tellurian, that by simply observing it, they can readily be understood. These motions play a very important part in the causes of the tides where they will again be considered.

The Tides

Owing to the erroneous theories generally given in text-books, the teacher as well as the learner has hitherto experienced great difficulty in clearly comprehending the causes of the tides. The old theory fails to give satisfactory explanation of the cause of the tides on the side of the Earth most remote from the Sun and Moon. According to the usually ascribed cause in our text-books, the side of the Earth farthest from the Sun or Moon is less influenced by the attractive force of these bodies, and consequently the solid portion of the Earth is drawn away, thus leaving the water bulged out behind, causing a tide. That such would be the fact is contrary to all known laws of physics. The attractive influence of the Sun or Moon will be less on the side of the Earth most remote from it, yet the gravitation of the Earth acting in the same line with those bodies, the water would be drawn more closely to the Earth in that part, if no other influence existed. That the attractive force is greatest at this point has been proved by experiment, it having been ascertained by actual test that a body weighs *more* at midnight than at any other hour.

If the old theory were correct, the weight of a body would be less at midnight than at any other hour, since at that hour the object is on the opposite side of the Earth from the Sun. Surely, if it were the absence of the attractive force which caused the water to bulge out at that point, the absence of the same force would be manifest in lessening the weight of an object at the same place. In accordance with the fact that an object weighs *most* when on the side of the Earth most remote from the Sun, it follows as a necessary consequence that when the object is on the side of the Earth nearest the Sun at midday, it would likewise *decrease* in weight.

A close examination of the movements of the Tellurian and a knowledge of the laws of planetary motion will present clearly to the mind a satisfactory theory of the tides, based upon known laws. If we attach a ball to the end of a string and revolve it around the hand, it will have a tendency to fly off at a tangent, or at a right angle to the string, and the faster it revolves the greater will be the tension on the string. This then, will illustrate the two great forces of planetary motion. The tendency that the ball, or body, will have to fly off is called the centrifugal force, and that which binds it to the center around which it revolves is called the centripetal force. If the ball suspended by the string be dipped in water and then removed, the water adhering to it will run to the lower portion of the ball by the force of gravity of the Earth acting upon it. Now, if we revolve the ball attached to the string, the water still adhering will fly off from the side opposite the string, that being the point on the ball farthest from the hand. In the movements of all heavenly bodies the centrifugal and centripetal forces are precisely equal, as may be illustrated by the string. The general law of matter, that all bodies attract all others in proportion to their mass, and inversely as the square of the distance increases, should be considered. We are now ready to understand the tides.

Solar Tides. The Earth revolves on its axis in twenty-four hours, and thus there are two solar tides, or tides caused by the Sun, each day, one beneath the Sun, and the other on the side of the Earth most remote.

These are constant and always remain in the same relative position to the Sun, thus always keeping an elevation of water directly under the Sun, and one on the side of the Earth opposite. Each particular portion of the Earth passes these two forces daily, and as the seas pass them, a tide is caused by the water being raised above its natural level on the Earth's surface. To the observer, who is necessarily moving with the Earth's revolution, these tides appear to be following each other around the globe, when, in fact, they always keep the same relative position to the Sun, and it is the revolving of the Earth on its axis that causes objects on its surface to pass the solar tides. On the portion of the earth nearest the Sun the water is drawn out or piled up by the gravitating force lodged in the central orb, as the particles of water move very easily among each other, and yield more readily to the attractive influence than do the solid portions of the Earth. The Solar tide most remote from the Sun is caused by the centrifugal force produced by the revolution of the Earth around the Sun. Thus, the bulging out of the water on one side is caused by the centripetal force, and on the other side by the centrifugal force. It must be remembered that every particle of the Earth feels the effect of these two forces in its orbital motions, but the surface of the Earth next to the Sun, being nearest, feels the centripetal force in excess of the centrifugal, and the surface most remote from the Sun feels the centrifugal force in excess of the centripetal. The effect is seen in the tide. At the center of the Earth these two forces are exactly equal. Again, the side of the Earth most remote from the Sun is 8,000 miles farther from the Sun than the side turned toward it, and, consequently, the farther side moves more rapidly, producing greater centrifugal force.

Lunar Tides. We have, so far, spoken only of the solar tide. By far the greater and more sensible tide is caused by the Moon. Although the Sun is vastly larger than the Moon, yet its effect in producing the tide is much less than that of the latter body, the ratio being about one to three. That is, while the Sun raises the tide one foot, the Moon will raise it three feet. This is owing to the close proximity of the Earth and Moon. The Earth, as a whole, feels the gravitating of the Earth and Moon. The Earth, as a whole, feels the gravitating influence of the Sun much more than that of the Moon, but upon a particular portion of the Earth the influence of the Moon is greater.

There are two lunar tides, one on the side of the Earth facing the Moon, caused by the centripetal force of the Moon, and the other on the side of the Earth most remote from the Moon, caused by the centrifugal force produced by the Earth revolving around the common center of gravity of these two bodies. The daily and yearly motions of the Earth are generally understood, and, although we are in the habit of considering the Moon as simply revolving around the Earth, it must be remembered that the attraction is mutual; that both bodies describe orbits about their common center of gravity, and that while the Moon obeys the attraction of the Earth, the latter equally follows that of the former, by which it is at every instant drawn from the path it would pursue if that influence did not exist. This motion is very perfectly shown by the Tellurian. The lunar tides traverse the globe with the Moon, and as to time differ from the solar tides. The centripetal and centrifugal forces, acting in oposite directions, will also cause tide waves on opposite sides of the Earth, and, as the forces are always equal, the tides produced must also be equal. The Earth and Moon are comparatively close to each other, and the Earth is much larger, and consequently the orbit of the Earth around the center of gravity will be very small; yet the centrifugal force does not depend altogether upon the velocity with which a body moves, but upon the size of the curve described by the body. It will be observed that the Earth in its lunar motion turns very short corners, as it were, and thus there will be a great tendency in every part of the Earth's lunar orbit to throw off, or bulge it out, as in the tides.

Spring Tide. If the Moon is in conjunction, as at new Moon, the centripetal forces of the

Sun and Moon act together, and the solar and lunar tides will occur at the same time, and, as shown in figure 1, a large tide will be the result. On the side of the Earth turned away from these bodies, an equal tide will occur; at that point the Earth feels the centrifugal force produced by its revolutions around the Sun, and its revolutions around the common center of gravity. These two forces act in one and the same straight line. Now let us place the Moon and Sun so as to be in opposition, or full Moon, and again there will be a large tide on the side of the Earth facing the Sun, caused by the centrifugal forces generated by the revolution of the Earth around the common center of gravity, and centripetal force of the Sun acting together. On the side of the Earth facing the Moon the centripetal force of the Moon will act in conjunction with the centrifugal influence produced by the revolution of the Earth around the Sun, and thus there will be equal tides on opposite sides of the Earth. These high tides are denominated Spring Tides.

Neap Tide. At first and third quarter, or when the Moon is in quadrature, as shown in Figure 2, the centripetal force of the Moon will act at right angles with the centripetal force of the Sun and the centrifugal force produced by the Earth revolving around the common center of gravity of the Earth and Moon, and its revolutions around the Sun, likewise act at right angles to each other, and we will have the phenomena of the solar and lunar tides about 90° apart, the solar tide being the smaller, and the lunar tide the larger. These are called Neap Tides. The tide now will neither rise so high, nor fall so low as at new or full Moon. The solar tides occur at regular intervals of twelve hours, but the Moon revolving eastward comes to a given meridian later and later each day and thus there will be a lunar tide on an average every twelve hours, and twenty-five minutes or fifty minutes later each day. The ponderous weight of such a great mass of water as constitutes a tide is not suddenly raised by the Moon's attraction, but yields slowly to the unseen force, *so that the greatest effect is noticed after the Moon has passed.* For this reason the highest point of the tide is, in the open ocean, 45° behind or east of that body.

The Heights of Tides in Different Parts of the World. The time is often essentially varied by local influences as the direction of coasts and the peculiar shape of bays and mouths of rivers. If a place is situated on a large bay, with but a narrow strait connecting it with the sea, the tide will be longer in rising, as the bay has to fill up through a narrow gate. Hence it is that at New York high tide usually occurs eight or nine hours after the Moon has passed the meridian. There are some very interesting phenomena to be noticed in the tide in Puget Sound, owing to the peculiar form and size of that body of water and its connection with the ocean by the Straits of Juan de Fuca. At Olympia, the extreme southern point of the Sound, the tide varies much, both as to height and time from the tide on the main coast near by. The tidal wave comes westward at the rate of about 1,000 miles per hour; therefore, in striking the eastern coast of continents the water rushes far up into bays and rivers, and oft-times with great violence, producing wonderful high tides. The water from the Atlantic Ocean is thus swept into the Gulf of Mexico, and raises a tide on the coast of Central America to the height of eighteen feet. The mouth of the Gulf is wide, and so shaped as to gather the water from a large part of the ocean, and thus concentrate the tide, as it were, in comparatively narrow limits at the western extremity of the Gulf. This phenomena is manifested more in the Bay of Fundy than in any other place on the globe, at which place it rises at times to the enormous height of seventy feet. By observing the map of the Bay of Fundy and its surroundings, and considering the shallowness of the Atlantic Ocean beyond, we may readily understand why the water would rush up the Bay with such force. The further up a bay or river, the later the tide will occur. At Cape Good Hope there is scarcely any perceptible tide, owing to the fact that as the water rushes against the lower portion of the eastern coast of Africa, it is, by the angle of the coast line, caused to flow down to the Cape by its own force. It reaches the Cape just as the water in that vicinity is going out so the change is not noticed and the water appears to remain unmoved at that particular place.

The Height of Tides at Different Seasons of the Year. At certain seasons of the year the tides occurring at night and during the day attain different heights—the greatest difference being in June and December. In March and September there will be no perceptible difference between the day and night tides. This phenomena is occasioned by the inclination of the Earth to the plane of the ecliptic. The tides must necessarily be under the lifting forces and directly opposite to those forces. If we place the Moon in conjunction with the Sun, in the beginning of winter, the lifting forces will be south of the equator, and the centrifugal force will be north of the equator on the opposite side from the centripetal forces. Thus in the Northern and Southern Hemispheres we will have each alternate tide higher. Now, change the bodies to like position at the beginning of summer, and just the opposite will be the result, and again each alternate tide will be higher. The same may be shown by placing the bodies in opposition. Thus the pupil may readily see the cause of this phenomenon. When the Moon is in perigee the tide will be somewhat increased, owing to the nearness of the bodies at that time. Likewise, when the Moon is in apogee, the tide will be slightly reduced. The character of tides are also affected by the winds. Strong winds may either retard or hasten the movements of the water, or may increase or diminish the height of the tide.

There is no perceptible tide in the Great Lakes, owing to the small compass and shallowness of the water.

We have spoken in this article of the increased weight of an object on that part of the earth most remote from the Sun, but this will in no wise conflict with the centrifugal force in producing the tide, as this force acts as a tangent to the orbit, or at right angles to the radius-vector.

Time

For the convenience of civil life, a specific measurement of time is essential. The great standard is the revolution of the Earth on its axis which is always the same. The transit of a heavenly body is the moment it is on the meridian of the observer; when it is on the meridian nearest the observer, it is the *upper transit;* when on the one most remote, the *lower transit.* The period of the revolution of the Earth on its axis is 23 houis and 56 minutes, the time between two successive upper transits of a star, which is called *siderial time.* This is always used for astronomical purposes, and in hours is numbered from 1 to 24. The right ascension, or siderial time of the Earth and Sun, is marked upon the chart for any day in the year. On March 21 siderial noon corresponds with solar noon, and on September 23 siderial noon to midnight.

The Earth in its annual motion around the ecliptic continually changes its longtitude, at the rate of about 1° each day; it must, therefore, make 1° more than a revolution on its axis between two successive upper transits of the Sun. This is called solar, or Sun, time. The difference, then, between, siderial and Sun time is about four minutes; that is, if the Sun and a star should make their upper transits at the same moment, the star would return to the same meridian about four minutes before the Sun. In a year the star would make one more upper transit than the Sun, or, in other words the siderial days in a year are one more than the solar days.

The Earth moves faster in its orbit when in perihelion than when at aphelion and thus varies the rate at which the Sun and Earth change their right ascension. The Sun would, therefore, make its upper transit at irregular intervals. As it will be seen, when the Earth is changing its right ascension faster, as at perihelion, it must be a little more than twenty-four hours between two successive upper transits of the Sun, and when it is in another part the time between two upper transits of the Sun will be a little less than twenty-four hours. At the equinox the Sun moves obliquely to the meridian, and also varies the solar noon. This time, as measured by the Sun, is *apparent time*. *Mean time* is the average of all the solar days in the year. This is the *civil day* of twenty-four hours, and begins at midnight, or when the Sun is on the lower transit. It is divided into two periods of twelve hours each, beginning at midnight and noon. The interval between *mean* and apparent time is called *equation of time*. The equation of time must sometimes be added to and sometimes subtracted from, apparent time, to give the mean. When the Sun is "slow" it must be added and when "fast" it must be subtracted to give the mean. The greatest additive value is about fifteen minutes, February 11, and the greatest subtractive value is about sixteen minutes, November 3. The equation of time is 0 four times a year—April 15, June 15, September 1 and December 24. The number of minutes the Sun is "slow" or "fast" for any day in the year is marked on the chart on the Tellurian, and the positions of the Earth relative to the Sun should be noticed in connection with the above explanations, when it will become very simple. As the Earth revolves to the east any place east of a given meridian must be later, and west, earlier.

The hour band is numbered from one to twelve, and then the same repeated, so one twelve will show the noon meridian and the other midnight. When it is desired to find the difference in longtitude or time between any two given places bring the meridian passing through one of the points to the noon twelve and notice where the meridian passing through the other point strikes the hour band, which will show the time. All problems in *Longtitude* and *Time* can thus be solved and simplified.

Precessson of the Equinoxes

By comparing the observations of the present with those mapped out by the astronomers of the last .2,000 years, it is found that the equinoxes are slightly shifting their places among the fixed stars; the change being about 1° towards the west in about seventy years. But this change is so slow that it is only by the refined instrument of modern times that it can be accurately determined. It will be noticed that the equinoxes are 90° from the poles, and a change in the equinoxes must necessarily be produced by a change in the direction of the poles. Precession is really produced by a very slow motion of the North Pole of the Earth around the pole of the ecliptic, requiring nearly 26,000 years for the pole to be carried all the way around, the motion being in an opposite direction to that of the Earth around the Sun. This motion is due mainly to the great attractive force of the Sun and Moon upon the protuberance of the Earth's equator and the centrifugal force of the Earth. At present the North Pole of the Earth has its greatest inclination from the Sun, in that part of the ecliptic denoted by the winter solstice, and is directed very nearly toward the pole star. As the North Pole swings to the right the Sun will appear in the equinoxian points about twenty minutes earlier each year. At this rate, in about 6,500 years the equinoxes will fall back in reference to the fixed stars to the present solsticial points and the solsticial points to the present position of the equinoxes. That is, the North Pole will have its greatest inclination from the Sun at that point that now marks the autumnal equinox. In about 6,500 years more the position of the solsticial and equinoxial points will be reversed and the North Pole will be directed about 46° from the present pole star, and so on, in a little less than 26,000 years the pole will have returned to its present position. The return of the Sun in reference to a fixed star is the *siderial* year and its return to the same equinox is called the *tropical* or *equinoxial* year. Their exact lengths are:

	Days.	Hours.	Min.	Sec.
Siderial Year	365	6	9	9
Tropical Year	365	5	48	46

The latter is used in the calendar as the return of the seasons depends upon it. In explaining the above with the Tellurian, move the Earth around in its orbit a number of times and the change in the direction of the poles will be observed, or throw the mechanism "out of gear" when at the winter solstice, December 21, and move the long arm backward slightly or as much as desired which will illustrate plainly the change of the direction of the Earth axis and the consequent *precession* of the *equinoxes*.

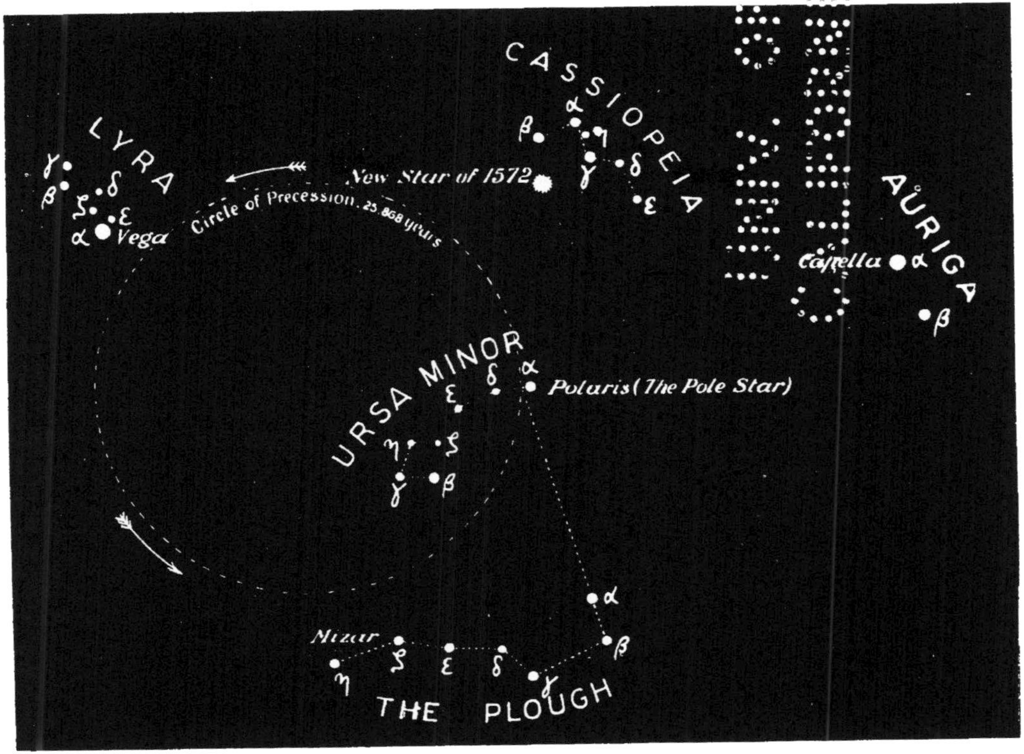

PLATE V.
Precession of the Equinoxes.

The Solar System

Our Solar System is composed of the Sun, the great central luminary, eight primary planets, twenty moons, about 200 known asteroids and a number of comets and meteors. The Sun is in the center of the system, around which all the bodies revolve in elliptical orbits. It is 860,000 miles in diameter with a density about one-fourth that of the Earth. The Sun is the source of all the light and heat of the Solar System except the dim twinkling that come from the far distant stars. The Sun is a molten mass of similar composition to our Earth, and revolves on its axis in about twenty-six days. The first planet from the Sun is Mercury, 35,750,000 miles distant, and is about 2,992 miles in diameter and makes a revolution around the Sun in 87.97 days. The second planet, Venus, makes a revolution around the Sun in 224.7 days, at a distance of 66,750,000 miles and is 7,660 miles in diameter. The third planet in order is our Earth, at a mean distance of 92,330,000 miles. It makes a complete revolution in 365 days, 6 hours, 9 minutes, and moves in its orbit at the rate of 68,000 miles per hour, or nearly 19 miles per second. Mars, the fourth planet, makes a revolution around the Sun in about 687 days at a mean distance of 141,999,000 miles and is 4,211 miles in diameter. It has two satellites, which revolve very rapidly around the planet, owing to their close proximity. Next beyond Mars is the region of asteroids, a large collection of small planets varying in size from a few miles to 100 miles in diameter. There have been about 300 discovered and named. The fifth planet is the great Jupiter, 86,000 miles in diameter. It makes a revolution around the Sun in 11.86 years, at a mean distance of 480,000,000 miles. Around this planet revolve four satellites at distances from 260,000 miles to over 1,000,000 miles from it. At certain seasons this planet, like Venus and Mars, shines with great brilliancy. Saturn is the sixth in order of distance from the Sun, around which it revolves in 29½ years, at a mean distance of about 880,000,000 miles, and is 70,500 miles in diameter. It has eight moons or satellites revolving around it. Two huge rings, a hundred miles in thickness, girdle this planet, the larger having an exterior diameter out of 169,000 miles. The seventh planet is Uranus, at a mean distance of 1,771,000,000 miles from the Sun. It is 31,700 miles in diameter and makes a revolution in 84 years, being accompanied by four satellites. At a distance of 2,775,000,000 miles from the central orb rolls the farthest planet of the system, Neptune. It makes a revolution in about 164¾ years, is 34,500 miles in diameter and is accompanied by one satellite.

A number of comets revolve around the Sun in very eccentric orbits and many pass far beyond the orbit of Neptune.

Glossary of Technical Terms Used in this Book

Aberation (a wandering away). Generally applied to a real or apparent deviation of the course of a ray of light.

Altitude. The apparent angular elevation of a body above the horizon, usually expressed in degrees and minutes. At the horizon the altitude is zero; at the zenith it is 90°.

Annular (ring shaped). Having the appearance or form of a ring.

Aphelion. The point of the orbit of a planet in which it is farthest from the Sun.

Apogee. The point of an orbit in which the Moon is farthest from the Earth. Applied only to the most distant part of the Moon's orbit.

Apsis (pl. apsides). The two points of an orbit which are nearest to, and farthest from, the center of motion, called, respectively, the lower and higher apsis. The line of apsides is that which joins these two points, and so forms the major of an ecliptic orbit.

Azimuth. The angular distance of a point of the horizon from the North or South. The azimuth of a horizontal line is its deviation from the true North and South direction. The azimuth of the East and West points is 90°.

Centrifugal (receding from the center). If a body be revolved about a center, the tendency it has when in motion to fly off tangent to the orbit in which it revolves, is called centrifugal force.

Centripetal (tending toward the center). The attractive force of a body drawing another body to it. The gravitating force of one body upon another.

Colures...The four principal meridians of the celestial sphere all of which pass from the pole, and one of which passes through the equinox, and one through each solstice. They mark the circles of 0 hr., 6 hr., 12 hr. and 18 hr. of right ascension, respectively.

Conjunction (a joining). The nearest approach of two heavenly bodies which seem to pass each other in their course. They are commonly considered as in conjunction when they have the same longtitude. The term is applied especially in the case of a planet and the Sun. The nearest approach is called superior conjunction when the planet is beyond the Sun; inferior when it is between us and the Sun.

Declination. The angular distance of a heavenly body from the Equator. When North of the Equator, it is said to be in North declination; otherwise, in South declination.

Digit. The twelfth part of the diameter of the Sun or Moon, formerly used to express the magnitude of eclipses.

Dip of the Horizon. At sea, the depression of the apparent horizon below the true level, owing to the height of the observer's eye above water.

Direct Motion. A motion from West to East among the Stars, like that of the Planets in general.

Ecliptic. The apparent path of the Sun among the Stars.

Ecliptic, plane of. The great plane extending through the center of the Earth and the center of the Sun.

Egress (a going forth). The end of the apparent transit of one body over another, when the former seems to leave the latter.

Equation of Time. The difference between the real and the mean Sun.

Equator. The great circle half way between the two poles of the Earth or heavens, dividing the Earth, or celestial sphere, into two hemispheres.

Equinox. Either of the two points in which the Sun, in its apparent annual course among the Stars, crosses the Equator. So called because the days and nights are equal when the Sun is at these points.

Hour Circle (see Meridian.

Inclination (to lean to). The inclination of an orbit is the angle it makes with a plane. The angle made with a perpendicular to a plane.

Ingress. The commencement of the transit of one body over the face of another.

Latitude. The angular distance of a heavenly body from the ecliptic, as declination is distance from the Equator.

Longtitude, celestial. The distance from the first point of Aries, measured East (the direction in which the Earth moves around the Sun), on the ecliptic; or from the point where the Sun appears at the Vernal Equinox.

Lunation. The period from one change of the Moon to the next. Its duration is 29½ days.

Mass, of a body. The quantity of matter contained in it, as measured by its weight at a given place. Mass differs from weight in that the latter is different in different places, even for the same body, depending on the intensity of gravity, whereas the *mass* of a body is necessarily the same everywhere.

Meridian, celestial. A great circle passing from the North to the South Pole, and crossing the Equator at right angles.

Nadir. The point of the celestial sphere directly beneath our feet, or the direction exactly downward.

Node. The point in which an orbit intersects the ecliptic, or other plane of reference. Usually applied to the Moon's orbit crossing the ecliptic.

Nutation. A very small oscillation of the direction of the Earth's axis. It arises from the fact that the forces which produce the precession of the equinoxes do not act uniformly, and may, therefore, be considered as the inequality of the precession arising from the inequality of the forces which produce it.

Oblate. Applied to a round body which differs from a sphere in being flattened at the Poles, as in the case of the Earth.

Obiquity of the Ecliptic. The inclination of the plane of the Equator to that of the ecliptic, which is equal to half the difference between the greatest meridian altitude of the Sun, which occurs about June 21, and the least, which occurs about December 21, which is 23½ degrees.

Opposition. A position of a planet with reference to the Sun. A planet is said to be in opposition when it is in the opposite direction from the Sun. The planet then arises at sunset and sets at sunrise. A superior planet only can be in opposition.

Orbit. The path described by a Planet around the Sun, or by a satellite around its primary Planet.

Parallels. Imaginary circles on the Earth, or in the heavens, parallel to the Equator, and having the Pole as their center.

Peri (near). A general prefix to denote the point at which a body revolving in an orbit comes nearest to its center of motion; as, perihelion, the point nearest the Sun; perigee, that nearest the Earth.

Precession of the Equinoxes. The Equator crosses the ecliptic at the First Point of Aries. The Precession of the Equinoxes consists of a backward movement of the First Point as a result of which in about 26,000 years, it will have traveled entirely around the ecliptic.

Quadrature. The position of the Moon when it is 90° from the Sun, and, therefore, in its first or last quarter.

Radius-vector. A straight line joining the Sun to a heavenly body.

Right Ascension. A term employed to locate a celestial body. It is the angular distance of that body from the First Point of

Aries as measured along the celestial equator to the east.

Siderial. Relating to the Stars. Siderial time is time measured by the apparent diurnal revolution of the stars. Each unit of siderial time is about 1-365th part shorter than a unit of mean time.

Signs of the Zodiac. The twelve equal parts into which the zodiac was divided by the ancient astronomers.

Solstices (standing points of the Sun.) Those points of the ecliptic where the Sun is most distant from the Equator, and through which the Sun passes about June 21 and December 21. So called because the Sun, having then attained its greatest declination, ceases its motion in declination and begins to return toward the Equator. The two solstices are designated as those of Summer and Winter respectively the first being in 6 hours and the second in 18 hours of right ascension.

Syzygy. The points of the Moon's orbit in which it is either New Moon or Full Moon. The line of the Syzygy is that which passes through these points, crossing the orbit of the Moon.

Transit (a passing across). The passage of an object across some fixed line, as the meridian, for example, or between the eye of an observer and an apparently larger object beyond, so that the nearer object appears on the face of the more distant one.

Umbra (a shadow). The darkest part of the shadow of an object where no part of the luminous object can be seen.

Vertical, angle of. The small angle by which the real direction of the Earth's center from any point on its surface differs from that which is directly downward, as indicated by the plumb-line.

Zenith. The point of the celestial sphere which is directly overhead.

Zodiac. A belt encircling the heavens on each side of the ecliptic, within which the larger Planets always remain. Its breadth is generally considered to be about sixteen degrees—eight degrees on each side of the eliptic.

How to Set Up and Adjust the Tellurian

To set up the Tellurian there is but one operation: Put Earth Arm (No. 3) into socket at (No. 7) and fasten with Set Screw (No. 8).

To adjust: Release the Clutch (No. 12); bring the Earth Arm over December 21 marked on the top of the stand. Revolve the Earth and the Moon on their common centers until the axis rod has its greatest inclination from the Sun. Fasten Clutch.

To adjust the Moon for any year: With Tellurian adjusted and in gear, revolve until Earth Arm is over the date of new Moon near est to January 1. Release Clutch and revolve Earth and Moon upon their common centers enough to bring the Moon Arm directly over the Earth Arm. This will never require more than one-half a revolution. Fasten the Clutch and the Tellurian will show all new Moon dates for that year or for several years.

To adjust for Eclipses: Bring to new Moon as above directed on date of a total or annual eclipse of the Sun as given by the almanac or other source. With the fingers, adjust the

Moon cam so that the Moon arm will be supported directly underneath one of the neutral marks on opposite sides of the cam. These are to represent the Moon's nodes. One should know whether it is at the ascending or descending node that the eclipse occurs. Then operate the Tellurian and all eclipses for that year or for a number of years may be foretold.

How to Demonstrate to a Class What the Axis of the Earth Means

In the past the teacher has used a pencil run through the center of an apple or an orange. The pencil represented the axis of the earth, the apple or orange, the earth revolving around the axis. Instead of this obsolete system, demonstrate with the Matlick Tellurian as follows: Simply draw the Long Pin (No. 14) from the top of the Earth Globe and reverse it, and place the part which protrudes from the top of the Earth Globe into the hole in which the Earth Globe revolves. The teacher will find that this represents the axis of the Earth, and by revolving the Long Arm (No. 3) around the Sun (No. 2) in the usual manner the axis of the Earth will be represented as continually pointing in the same objective direction inclined 23½ degrees from a perpendicular. In this demonstration, of course, the Earth Globe is removed, also the Time Band (No. 13).

Spotted something that isn't quite right?

Let us know and we'll fix it for future reprints. And as a thank you for your diligence, we'll send you a free Sothis book of your choice.

publisher@sothis.press

This page intentionally left blank

Made in United States
Orlando, FL
06 November 2026

90301614R00072